*How the book is laid out* – the many seaweeds and marine creatures are shown as the example below

## Beadlet Ane[...]

*Actinea equina*

**HABITAT** Upper, midd[...]
**WHERE TO LOOK** Wha[...]
  restricted to a cert[...]
  together with its a[...]
**DESCRIPTION** A shor[...]
  of the species is g[...]
  behaviour are also[...]
  similar in appeara[...]
  Most species have[...]
  relevant feature to aid identification.

**SEAWEEDS** (*see pp.4–17*) are marine algae and photosynthesise to produce their food but also absorb nutrients directly from the sea through their fronds. They are anchored to rocks by a holdfast which looks like roots but, unlike roots it does not contribute to the seaweed's growth. Seaweeds are divided into three groups, green, brown and red seaweeds. There are about 120 species found along Somerset's coast.

**SPONGES** (*see pp.17–18*) are very simple immobile animals , most of which form encrusting mats on rock and boulders in the intertidal zone. They feed by sucking water through their bodies and filtering out microscopic food items, including bacteria. There are several very similar encrusting species that are hard to tell apart along our coast, and probably many more to identify.

**JELLYFISH** (*see pp.19–20*) are soft-bodied animals with a domed umbrella shape and a central downward-facing mouth surrounded by tentacles, not all have a powerful sting. Although they appear to be completely at the mercy of tides and currents they can swim actively by means of pulsating contractions.

**SEA ANEMONES** (*see pp.20–22*) are animals closely related to corals. Unlike corals, which are sedentary in their calcareous rock-like home, anemones can move around, often to ensure they are in good areas to catch prey. They catch their prey (small fish, shrimps and small crabs) with their stinging tentacles.

**HYDROIDS** (*see pp.23–24*) are colonial animals, some with a leathery body structure. Many hydroids have rather fragile-looking, bushy or feathery forms as in the Sea Fir, but a few such as Oaten Pipes Hydroid have a robust leathery look about them. They all have small tentacles, some with a powerful sting, to catch food.

**MARINE WORMS** (*see pp.25–28*) fall into four basic types. Some such as the Spiral Worm build a structure by secreting calcium to form a tube in which to live, some, like the Sand Mason Worm use sand to build a solid tube to occupy. Others like the Blow Lug Worm live in sand in a burrow, and some like the Greenleaf Worm are active hunters and live out in the open.

**CRUSTACEANS** (*see pp.29–37*) have a hard, armoured, outer shell which has to be shed periodically to allow them to grow. They include the familiar crabs, prawns and barnacles. Many of the "dead" crabs you find on the tideline are the discarded shells of crabs still living happily on the beach. Barnacles are in effect small crab-like creatures lying on their backs and glued to the rocks. Their shell is modified to create a tough defensive wall that encircles them. When the tide covers them they use their feathery feet to capture very small particles of food.

**BRYOZOANS** (*see pp.38–42*) are colonial animals that grow mainly as flat encrusting mats on rocks or seaweeds, or form a substantial free-standing "body" as in Sea Chervil. The individual animals, called zooids, within the colony are packed tightly together. Each zooid has feeding tentacles that capture minute particles of food.

**MARINE MOLLUSCS** (*see pp.42–63*) can be divided into three groups: gasteropods – those with one shell such as whelks and winkles, and sea slugs which have a single shell hidden internally; bivalves which have two hinged shells like the familiar cockles; and cephalopods which includes octopuses, squids and cuttlefish which have an internal shell. Many of the marine molluscs found living in Somerset's sea may have their empty shells washed up along the tide line. A particularly good time to hunt for shells is following stormy periods when shells lying in offshore hollows may be brought inshore by extra powerful wave action.

**ECHINODERMS** (*see pp.63–65*) are animals with radial symmetry, rounded bodies and hard, often spiny, skin and include starfish and urchins. One echinoderm that can be found on our coast, the **Green Sea Urchin** ***Psammechinus miliaris***, has the legs fused into a rigid case covered with spines. They are all active hunters feeding on things such as marine molluscs, colonial animals such as ascidians, worms and carrion.

**ASCIDIANS (SEA SQUIRTS)** (*see pp.65–66*) are animals with rather jelly-like bodies. They are tiny and live mainly in colonies. The individuals have a well-defined body structure but the colony often has no set shape. They feed by pumping water through their bodies and filtering out food particles. In the Star Ascidian each animal has a mouth but they all share a single central exhalant hole where the waste water is ejected.

**FISH** (*see pp.66–71*) are divided into two groups, those with a cartilaginous skeleton (sharks and rays), and those with a bony skeleton (all other fish). Sharks and rays produce eggs in tough leathery cases, known as mermaid's purses. They are laid well offshore and deliberately entangled in seaweeds or wedged amongst boulders. Each egg-case holds a single egg which may take several months to mature before hatching. Most egg-cases that you find will be empty and successfully hatched or will have been predated.

Every Somerset beach will have some great wildlife to be found there, but they differ from east (around Brean) to west (around Porlock). This is because Somerset's coast is also the estuary of the River Severn. It is more salty in the west and less salty in the east. Our marine life is adapted to live in salty water and the more east you go, and up river, the less they like it. In consequence the number of species you are likely to find decreases as you go east.

# Where to look

### SAND AND MUDFLAT BEACHES
(like at Brean and Berrow) – Expect to find large numbers of worms living in the sand. Also home to many bivalve (two-shelled) molluscs which live shallowly buried in the sand.

### SHINGLE BEACHES
(like at Stolford and Porlock) – This a tough place to live. As the sea washes across the shingle the stones are constantly moved about, grinding and squashing things that might try to live amongst the stones. This is still a good place to hunt along the tideline for marine life washed ashore.

### BOULDER BEACHES
(like those exposed at low tide at Minehead and Porlock Weir) – The boulders here are too big to be moved around much by the tides and many species will be found living on and under them such as sponges, colonial animals and crabs.

### ROCKY BEACHES
(like at Watchet and Kilve) – Rocky beaches give a solid base for seaweeds to grow, which in turn provides habitat and food for large numbers of marine molluscs. Rocky beaches also tend to have rockpools providing places for sea anemones, fish and crabs to live whilst the tide is out.

# How to look

**KEEPING SAFE** – **Firstly you need to be looking safely!** This coast has the second highest tidal range in the world. That means the tide goes out a very long way exposing large tempting areas to explore. But it also comes back in very quickly across our shallow shelving beaches. You should only attempt to explore low tide areas by following the tide as it goes out and know what time the tide is expected to come back in. You can easily find out what the tides are doing by buying a copy of a local tide times booklet from cafes and shops along our coast. There are several websites that give free seven-day tidal predictions. It is also strongly recommended that you **do not go low tide exploring on your own**. Take someone to keep an eye on the tide whilst you concentrate hunting in those rockpools.

**STRANDLINES** – The strandline should be your first port of call as you go on the beach. This will often give a great indication of what might be living between the tides. And if the tide is right in when you arrive on the shore, hiding most of the beach, the tidelines can still give you the chance to learn something about what might be living beneath the waves on the lower shore.

**TOP TIPS FOR FINDING THINGS** – Seaweeds are worth looking at in their own right but it is well worth gently turning seaweeds over to see what might be living beneath them. Looking under rocks and boulders can be very worthwhile. Many small fish and crabs will be found by gently lifting big rocks. Also the underside of rocks and boulders may have starfish clinging to them and many colonial animals like sea squirts, bryozoans and sponges encrust the undersides of rocks. But please do this very gently, and put rocks back very carefully as you found them, not damaging the wonderful wildlife that lives in these hidden places. Don't rush when looking! A lot of rockpool wildlife hide under stones – animals will have hidden themselves quickly away when you approached. Spend time watching, keeping very still, fish and crabs will often move back out from a hidden crevice if you give them time. Overhanging rocks that create small cave-like places are often a favourite place for sea anemones, sea squirts and starfish to live. Take a torch so you can have a better look at what is hidden in those dark places.

    The tides of the sea on our beaches do not always go up and down to the same level. When the sun and the moon are in line with the Earth their gravitational pull will give us very low tides and very high tides. At other times we get average up and down tides. Also you need to remember that the low and high tide are an hour later each day. Very low tides, known as spring tides, come round regularly every month. Spring tides are when the sea drops to the very lowest point on the beach and this can be a great chance to explore for marine wildlife not often seen. Spring tides are your best chance to look for things such as starfish, squid eggs and sponges.

# Common Green Branched Weed
*Cladophra rupestris*

**HABITAT** Middle and lower shore
**WHERE TO LOOK** On rocks and boulders and in
  rockpools. All beaches. Common.
**DESCRIPTION** A green seaweed. Usually up to 20cm
  long but can reach 30cm. Dark green, sometimes
  grey/blue. Forms fine filamentous bunches. Feels
  stiff and wiry to the touch.

Detail of filamentous frond

# Sea Lettuce
*Ulva lactuca*

**HABITAT** Upper, middle and lower shore.
**WHERE TO LOOK** On rocks and boulders and in
  rockpools. All beaches. Very common.
**DESCRIPTION** A green seaweed. Up to 100cm long.
  Pale to dark green and slightly shiny. The frond is
  very thin and flat and may have unevenly toothed
  or lobed margins. One of several similar species.

Frond with unevenly lobed margin

# Gutweed
*Ulva intestinalis*

**HABITAT** Upper and middle shore.
**WHERE TO LOOK** On rocks, boulders and stones, in rockpools and also on shells. All beaches. Very common.
**DESCRIPTION** A green seaweed. Up to 30cm long. Pale green and rather slimy looking. Unbranched, each frond is tubular and slightly inflated. One of several similar species.

Detail of fronds

5

# Oarweed (Kelp)
*Laminaria digitaria*

**HABITAT** Lower shore.
**WHERE TO LOOK** On rocks and boulders, Porlock and west. Only common at Gore Point.
**DESCRIPTION** A brown seaweed. Up to 150cm long. Brown, tough and leathery with a broad blade divided into fingers. Stem oval, smooth and shiny, can be bent double without breaking. **Forest Kelp Laminaria hyperboria** also occurs here. The circular stem is very rough and will snap if bent double.

Stem is oval in cross-section

# Sugar Kelp
## *Saccharina latissima*

**HABITAT** Lower shore.
**WHERE TO LOOK** On rocks and boulders and in rockpools. Minehead and west. Uncommon.
**DESCRIPTION** A brown seaweed. Up to 150cm long. Brown to golden-brown with an uneven surface and with distinctive crinkly edges.

Detail of crinkly edged frond

# Egg Wrack
## *Ascophyllum nodosum*

**HABITAT** Middle shore.
**WHERE TO LOOK** On rocks and boulders. All beaches. Very common.
**DESCRIPTION** A brown seaweed. Up to 150 cm long. Olive green to dark brown and distinctively thin and knobbly. Fronds are 10mm wide with large egg-shaped air bladders. It often has the small tufted, red seaweed Wrack Siphon Weed growing on it.

Wrack Siphon Weed growing on it

# Channelled Wrack
*Pelvetia canaliculata*

**HABITAT** Upper shore.
**WHERE TO LOOK** On rocks and boulders. All beaches.
Very common.
**DESCRIPTION** A brown seaweed. Up to 15cm long.
Olive green to dark brown. Grows in short forked
tufts. It has a distinctive channel or gutter along
the frond.

Fronds have distinctive channels

# Serrated or Toothed Wrack
*Fucus serratus*

**HABITAT** Middle and lower shore.
**WHERE TO LOOK** On rocks and boulders. All beaches.
Very common.
**DESCRIPTION** A brown seaweed. Up to 60cm long.
Golden brown to dark olive green. Has flattened
fronds with saw-toothed or serrated edges. There
are no bladders.

Fronds have serrated edges

# Spiral Wrack
*Fucus spiralis*

**HABITAT** Upper shore.
**WHERE TO LOOK** On rocks and boulders. All beaches. Common.
**DESCRIPTION** A brown seaweed. Up to 20cm long. Olive green to brown. The fronds are short and flat with a distinctive central rib and often have a spiral twist. Does not have bladders along the stem like Bladder Wrack but may have swollen reproductive structures at the tips of the fronds.

Swollen reproducive structures at tips

# Bladder Wrack
*Fucus vesiculosus*

**HABITAT** Middle shore.
**WHERE TO LOOK** On rocks and boulders. All beaches. Very common.
**DESCRIPTION** A brown seaweed. Up to 80cm long. Olive green to brown. Flat and branched, with distinctive bladders, usually in pairs, either side of a central rib along the frond.

Distinctive bladders

# Irish Moss
## *Chondrus crispus*

**HABITAT** Middle and lower shore.
**WHERE TO LOOK** On rocks, boulders and in rockpools. Lilstock and west. Common.
**DESCRIPTION** A red seaweed. Up to 20cm long. Red/brown or dull purple and often with the tips of fronds bleached green/yellow. Forms short, bushy, branching clumps.

A short, bushy, branching clump

# Grape Pip Weed
## *Mastocarpus stellatus*

**HABITAT** Lower shore.
**WHERE TO LOOK** On rocks, boulders and in rockpools. Porlock and west. Uncommon.
**DESCRIPTION** A red seaweed. Up to 10cm long. Brown/purple, forming straggly clumps. The fronds are channelled at their base, then flatten out and start repeatedly branching. Only female plants have the pip-like reproductive structures dotting the flat parts of the fronds.

Pip-like reproducive structures

## Bunny Ears
*Lomentaria articulata*

**HABITAT** Lower shore.
**WHERE TO LOOK** On rocks, boulders and in
rockpools. Porlock and west. Uncommon.
**DESCRIPTION** A red seaweed. Up to 10cm long.
Shiny red/orange and in branching clumps. The
fronds are small and cylindrical and with regular
constrictions making them look rather like
strings of sausages.

Fronds have regular contrictions

## Red Fringed Weed
*Calliblepharis ciliata*

**HABITAT** Lower shore.
**WHERE TO LOOK** On rocks, boulders and in
rockpools. Porlock and west. Uncommon.
**DESCRIPTION** A red seaweed. Up to 30cm long. Red
or red/brown. Flat, oval branching fronds, with
distinctive eyelash-like fringing.

Fronds have eyelash-like fringing

# Red Rags
*Dilsea carnosa*

**HABITAT** Lower shore.
**WHERE TO LOOK** On rocks, boulders and in rockpools. Porlock and west. Uncommon.
**DESCRIPTION** A red seaweed. Up to 20cm long and 25cm wide. Red/brown with a thick, almost plastic-like, feel to the fronds. Grows in an irregular branched way with broad lobes, often with holes in the fronds.

Fronds feel plastic-like

# Coral Weed
*Corallina officinalis*

**HABITAT** Middle and lower shore.
**WHERE TO LOOK** On rocks, boulders and in rockpools. Lilstock and west. Very common.
**DESCRIPTION** A red seaweed. Up to 6cm long. Pale pink and stiff with flattened, branching fronds. Often found along the tideline dead and bleached white.

Dead frond bleached white

## Common Pale Pink Paint Weed
*Lithophyllum incrustans*

Encrusting a shell

**HABITAT** Middle and lower shore.
**WHERE TO LOOK** On rocks and boulders in rockpools and on shells. All beaches. Very common.
**DESCRIPTION** A red seaweed. Irregular-shaped patches up to 10cm across. Pale to dark pink, forming thin hard, sometimes knobbly, encrusting patches on rocks, stems of brown seaweeds and sometimes shells. Difficult to tell apart from several similar species that occur along this coast.

## Red Siphon Weed
*Heterosiphonia plumosa*

Detail of feathery frond

**HABITAT** Lower shore.
**WHERE TO LOOK** On rocks and boulders, sometimes on the stipes of large seaweeds like Kelp. Porlock and west. Uncommon.
**DESCRIPTION** A red seaweed. Grows to 15cm. Rich red and quite tough in texture, sometimes slightly translucent. The fronds are flat, feathery, and branching.

# Creeping Chain Weed
*Catenella caespitosa*

**HABITAT** Upper and middle shore.
**WHERE TO LOOK** On shaded rocks and overhangs and under brown seaweeds. Brean to Blue Anchor. Common.
**DESCRIPTION** A red seaweed. Fronds to 20mm long and 2mm wide, but forming patches to 100cm across, sometimes more. Brown-purple, moss-like and mat-forming, with tiny, very narrow, creeping fronds.

Tiny creeping fronds form patches

# Creephorn
*Chondracanthus acicularis*

**HABITAT** Lower shore.
**WHERE TO LOOK** On rocks, boulders and in rockpools. Minehead and west. Common.
**DESCRIPTION** A red seaweed. Fronds to 10–12cm long and 1–2mm wide, forming mounds up to 20cm across. Reddish-purple turning yellow/green in late summer. Fronds are cylindrical and slightly flattened and are multi branched, forming an entangled mass.

Detail of multibranching fronds

# Dumont's Tubular Weed
*Dumontia contorta*

**HABITAT** Lower shore.
**WHERE TO LOOK** On rocks and in rockpools. Lilstock and west. Common.
**DESCRIPTION** A red seaweed. Up to 30cm long. Dark brown/red, soft and slimy. Very long and narrow. Fronds are about 5–8mm wide and tubular. Branches at irregular intervals. Often looks rather dirty as sand sticks to it.

Detail of fronds

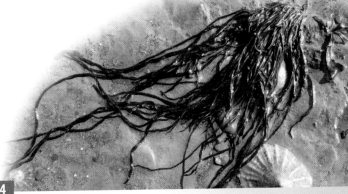

# Dulse
*Palmaria palmata*

**HABITAT** Lower shore.
**WHERE TO LOOK** On rocks and boulders. Watchet and west. Common.
**DESCRIPTION** A red seaweed. Up to 50cm long. Purplish-red and leathery looking. Fronds are up to 80mm wide, often simple and strap-like, but can be forked and branching.

Detail of frond

# Pepper Dulse
*Osmundea pinnatifida*

**HABITAT** Middle and lower shore.
**WHERE TO LOOK** On rocks and boulders. Porlock and west. Common.
**DESCRIPTION** A red seaweed. Up to 10cm long. Brown purplish-red, when growing higher up the shore can be green/yellow. Short tufted fronds are fern-like and flattened. Sometimes forms extensive carpets over rocks. Tastes peppery. The similar ***Osmundea ramosissima*** and **False Pepper Dulse** ***Osmundea hybrida*** have also been recorded.

Short fern-ilke frond

# Wrack Siphon Weed
*Polysiphonia lanosa*

**HABITAT** Middle shore.
**WHERE TO LOOK** Only found growing on Egg Wrack (see p.6) as an epiphyte (i.e. without causing harm). All beaches. Common.
**DESCRIPTION** A red seaweed. Up to 10cm long. Purple/brown. Grows as small tufted, filamentous clumps on Egg Wrack.

Clumps grow on Egg Wrack

# Sea Beech
*Delesseria sanguinea*

**HABITAT** Lower shore.
**WHERE TO LOOK** On rocks and boulders. Minehead and west. Uncommon.
**DESCRIPTION** A red seaweed. Up to 20cm long. Rich red. Grows with several flattened lance-shaped fronds from the holdfast. The midrib of the frond is thick and tough.

Frond has a thick, tough midrib

# Purple Laver
*Porphyra purpurea*

**HABITAT** Middle and lower shore.
**WHERE TO LOOK** On rocks and boulders. All beaches. Common.
**DESCRIPTION** A red seaweed. Up to 50cm long. Reddish-brown. Forms very thin, shiny gelatinous sheets. The very similar, and hard to tell apart, **Tough Laver** *Porphyra umbilicus* also occurs on this coast.

Thin gelatinous sheet

# Purple Hair Weed
# (Purple Velvet Weed)
*Rhodochorton purpureum*

Velvet-like texture of mat

**HABITAT** Upper shore.
**WHERE TO LOOK** On rocks, in shallow sea-washed caves and in cavities amongst big boulders. Brean Down, Lilstock and Glenthorne beach. Scarce.
**DESCRIPTION** A red seaweed. Extensive mats up to 100cm. Rich purple-red with a velvet-like texture. Forms extensive mats covering cave walls.

# Shiny Orange/
# Yellow Green Sponge
*Protosuberites denhartogi*

Subsurface has canal patterns

**HABITAT** Lower shore.
**WHERE TO LOOK** Under rocks and boulders. Minehead and west. Uncommon.
**DESCRIPTION** A sponge forming patches to 10cm across. Yellow/orange sometimes green and smooth, rather shiny looking. Only about 1mm thick. Subsurface canal patterns, part of the feeding structures, may be seen if you look closely in good light.

# Breadcrumb Sponge
*Halichondria panicea*

**HABITAT** Lower shore.
**WHERE TO LOOK** On rocks and boulders, on and under boulders and also in crevices. All beaches. Common.
**DESCRIPTION** A mat-forming sponge to 30cm wide (in sheltered conditions), much bigger offshore. Dull yellow/orange, grey/white and green. The mats, sometimes with a more spiky appearance, have volcano-like oscula (holes), which are part of its feeding system.

Broken piece on the tideline

# Mermaid's Glove Sponge
*Haliclonia oculata*

**HABITAT** Lower shore.
**WHERE TO LOOK** Found attached to rocks when alive, but unlikely, as this is a subtidal species. Broken-off pieces can be found at the lowest point of the tide. Minehead and west. Uncommon.
**DESCRIPTION** A sponge with branched finger-like growths. Height to 20cm. Orange/yellow when alive, broken-off pieces bleached fawn/white. Look along the fingers for the oscules.

Look for oscules along the fingers

# Compass Jellyfish
*Chrysaora hysocella*

**HABITAT** Close to shore and pelagic.
**WHERE TO LOOK** In waters off all beaches and washed up on the tideline. Uncommon.
**DESCRIPTION** A jellyfish up to 30cm in diameter. Fawn-brownish bell with darker brown markings in a symmetrical pattern. Tentacles up to 100cm and can cause a nasty sting.

Washed up on the beach

# Blue Jellyfish
*Cyanea lamarckii*

**HABITAT** Close to shore and pelagic.
**WHERE TO LOOK** In waters off all beaches and washed up on the tideline. Uncommon.
**DESCRIPTION** A jellyfish up to 30cm in diameter, but smaller ones are much more common along our coast. The bell is a translucent blue/purple. Tentacles are quite long, up to 150cm. Can inflict a nasty sting.

Washed up on the beach

# Moon Jellyfish
*Aurelia aurita*

Washed up on the tideline

**HABITAT** Close to shore and pelagic.
**WHERE TO LOOK** In waters off all beaches and
   washed up on the tideline. Common.
**DESCRIPTION** The commonest jellyfish on our coast.
   Up to 25cm diameter. Translucent with four
   conspicuous circular pink/purple reproductive
   organs. Very small tentacles. Doesn't have a sting
   to worry about.

# Beadlet Anemone
*Actinea equina*

Out of water with tentacles drawn in

**HABITAT** Middle and lower shore.
**WHERE TO LOOK** On rocks, in rockpools, under rock
   overhangs and under large boulders. All beaches.
   Common.
**DESCRIPTION** Diameter up to 5cm. Usually coloured
   a dark red but can also be orange or green. The
   name beadlet comes from the blue,
   beadlike, stinging organs that
   encircle the column
   just below the
   tentacles.

# Strawberry Anemone
*Actinea fragacea*

Out of water with tentacles drawn in

**HABITAT** Lower shore.
**WHERE TO LOOK** On rocks, under rock overhangs and under large boulders. Hurlestone Point and west. Common.
**DESCRIPTION** Diameter up to 8cm. At first glance similar to the Beadlet Anemone but grows larger than that species, and the green dots on the column are diagnostic.

# Snakelocks Anemone
*Anemonia viridis*

Out of water

**HABITAT** Lower shore.
**WHERE TO LOOK** In rockpools and under large boulders and rock overhangs. Watchet and west. Common.
**DESCRIPTION** Diameter up to 6cm. Tentacles are usually green with purple tips, but can be a pale brown or grey. The green Snakelocks have algae living within their tissues in a symbiotic relationship, causing the strong green colour.

## Dahlia Anemone
*Urticina felina*

Column is hidden in sand and gravel

**HABITAT** Lower shore.
**WHERE TO LOOK** In rockpools and amongst
   boulders in gravel and sand. Hurlestone Point
   and west. Uncommon.
**DESCRIPTION** Diameter up to 15cm. Tentacles are
   usually pink and white banded, can also be pale
   orange. Column is hidden in sand and gravel.

## Daisy Anemone
*Cereus pendunculatus*

Brown form partly covered by sand

**HABITAT** Lower shore.
**WHERE TO LOOK** In sand and gravel amongst
   boulders and in rockpools. Minehead and west.
   Uncommon.
**DESCRIPTION** Diameter up to 10cm. Tentacles are
   mottled brown and dirty white, sometimes blue/
   grey. The column is buried in sand or gravel and
   the tentacles may often be partly
   covered by sand.

# By-the-wind-sailor
*Velella velella*

**HABITAT** Pelagic, lives out on the open ocean, floats on the surface.

**WHERE TO LOOK** Could be found on any beach, but most often Minehead and west. Can be common far out at sea, and most likely to be found washed ashore in the autumn sometimes in large numbers.

**DESCRIPTION** Blue/black, partly translucent hydroid (to 8cm long). Translucent fin on the upper surface and short blue tentacles hanging below.

Top view of upper surface

# Oaten Pipes Hydroid
*Tubularia indivisa*

**HABITAT** Lower shore.

**WHERE TO LOOK** On rocks and boulders. Minehead and west. Common.

**DESCRIPTION** This hydroid (to 15cm long) forms distinctive pale brown tubes which are hard but flexible. Colonies grow in clusters of tubes which are anchored on rocks or boulders. The feeding tentacles are pinky/orange and will only emerge when the tide is in.

Feeding tentacle

# Sea Fir
*Bugula plumosa*

Detail of plumes

**HABITAT** Lower shore.
**WHERE TO LOOK** On rocks. Minehead and west.
Common.
**DESCRIPTION** Pale fawn, bushy hydroid (to 8cm long) with distinctive spiral plumes. This species is very soft and collapses when out of water making it difficult to identify unless it is floating in water.

# Sea Fir
*Sertularia argentea*

Detail of branches

**HABITAT** Lower shore.
**WHERE TO LOOK** On rocks. Grows subtidal but quite often found washed up along the tideline after winter storms. Watchet and west. Common.
**DESCRIPTION** Pale fawn, rather straggly-looking hydroid (to 20cm long) with alternating short side branches.

# Sand Mason Worm
*Lanice conchilega*

**HABITAT** Lower shore.
**WHERE TO LOOK** On sand beaches and also in sand patches between rocks and boulders. Watchet Point and west. Common.
**DESCRIPTION** This worm is a filter-feeder on plankton. It constructs a distinctive tube (to 5cm in height) from sand, with a fan-like structure at the top designed to collect food particles. The worm itself is up to 20cm long but is always hidden down in the sand.

The distinctive tube with a fan-like top

# Blow Lug Worm
*Arenicola marina*

**HABITAT** Middle and lower shore.
**WHERE TO LOOK** On sand. All beaches. Very common.
**DESCRIPTION** The pit and adjacent pile of coiled sand is a familiar site on all sandy beaches. The pinky-red worm (to 20cm long) lives in a U-shaped burrow in the sand, with its mouth beneath the pit. A filter-feeder on plankton, it sucks in water, along with microscopic particles of food and sand particles. The waste material is ejected at the other end, forming the neat piles.

Pit with a coil of waste material

# Ragworm
*Perinereis cultrifera*

Detail of head

**HABITAT** Lower shore.
**WHERE TO LOOK** On rocks, boulders and seaweed. All beaches. Common.
**DESCRIPTION** A mainly carnivorous worm to 20cm long, feeding on small crustaceans and worms. Pale green/brown with a distinctive red dorsal blood vessel running the length of its body. Has a rounded and firm body. Lives under rocks and boulders and amongst seaweed holdfasts.

# Greenleaf Worm
*Eulalia viridis*

Sticky green egg sack

**HABITAT** Lower shore.
**WHERE TO LOOK** On rocks and under boulders and amongst the holdfasts of larger seaweeds. Minehead and west. Uncommon.
**DESCRIPTION** A bright green and very active worm to 10cm long. A scavenger feeding on small dead or dying animals. Sometimes seen hunting over rocks and boulders as the tide drops. Its green and sticky egg sacks may be found washed up along the tideline.

# Scale Worm
*Hormothoe imbricata*

**HABITAT** Middle and lower shore.
**WHERE TO LOOK** Under rocks and boulders and amongst seaweed holdfasts. All beaches. Common.
**DESCRIPTION** A mainly carnivorous worm (to 45mm long) feeding on very small crustaceans and worms. Flattened and slightly hairy looking with obvious overlapping scales. Underside is a shiny silvery-white. One of several similar scale worms found on our beaches.

Shiny silvery-white underside

# Spiral Worm
*Spirorbis spirorbis*

**HABITAT** Middle and lower shore.
**WHERE TO LOOK** On rocks and boulders, also found on shells and crab carapaces and the holdfasts of larger seaweeds. All beaches. Very common.
**DESCRIPTION** This tiny worm, up to 4mm wide, secretes a dirty white calcified spiral tube that is tightly glued to rock and boulders etc. Filter feeder on microscopic plankton, its feeding tentacles project from the shell when submerged.

Detail of calcified spiral tubes

# Keel Worm
*Pomatoceros triqueter*

**HABITAT** Lower shore and subtidal.
**WHERE TO LOOK** On rocks and boulders. Watchet and west. Very common.
**DESCRIPTION** This small worm secretes a grey/white, triangular (in cross-section) tube (up to 25mm long), that increases in width as the worm matures. A filter feeder on plankton, when the tide is in its feathery, orange feeding tentacles emerge.

The tube is triangular in cross-section

# Honeycomb Worm or Reef Building Worm
*Sabellaria alveolata*

**HABITAT** Lower shore.
**WHERE TO LOOK** On rocks or boulders at the lowest point of the tide. All beaches. Common.
**DESCRIPTION** Forms distinctive reefs (which can cover hundreds of square metres) composed of tightly packed tubes (worms and tubes to 40mm) made by the worms from sand and sediment. The worms are filter feeders on microscopic plankton.

Tightly packed tubes

# Sea Slater
*Ligea oceanica*

**HABITAT** Upper shore.
**WHERE TO LOOK** On cliffs, rocks and boulders and
under seaweed. All beaches. Common.
**DESCRIPTION** A mainly nocturnal, matt grey, armour-
plated animal (up to 3cm long) with distinctive
large antennae. An air breather but can stand short
periods of immersion. It feeds on rotting seaweed
and small dead marine creatures. The smaller
(to 12mm) **Pill Woodlouse**
***Armadillidium vulgare***
is also found on
the upper
shore.

The Sea Slater is large, up to 3cm long

# Beach Hopper
*Orchestia gammarellus*

**HABITAT** Upper shore.
**WHERE TO LOOK** Under strandline seaweed and
strandline debris. All beaches. Very common.
**DESCRIPTION** Small, to 18mm long, brownish and
shiny, with black eyes. Hops wildly when disturbed
and lands on its side. It is an omnivore, feeding on
rotting seaweed and small dead marine creatures.
There are a number of other similar beach
and sand hoppers on these
beaches. You will need
a hand lens, and
patience, to
sort them
out.

Detail of head, showing black eye

# Edible Crab
*Cancer pagurus*

**HABITAT** Lower shore.
**WHERE TO LOOK** Under rocks and boulders, often half-buried in gravel/sand. St Audries Bay and west. Common.
**DESCRIPTION** Distinctive pink/brown "pie crust"-shaped carapace (to 20cm wide) and the powerful black-tipped pincers are diagnostic. Only young crabs up to 6–8cm and aged around 3–4 years are usually found on the lower shore. Larger, mature crabs live out in deeper water. Living to 100 years, they catch small live prey including other crabs but also feed on carrion.

# Green Shore Crab
*Carcinus maenas*

**HABITAT** Middle and lower shore.
**WHERE TO LOOK** Buried in mud, in rockpools, under seaweed, rocks and boulders. All beaches. Very common.
**DESCRIPTION** This crab (carapace to 10cm wide) is very variable in colour, from green to red/brown and sometimes almost black with dull yellow blotches. Very small crabs are more variable in colour including white and orange. Feeds on worms, molluscs small fish and carrion.

Very small whitish-coloured crab

# Hairy Crab
*Pilumnus hirtellus*

**HABITAT** Lower shore.
**WHERE TO LOOK** In small holes and crevices in
rockpools and under boulders and seaweeds.
Watchet and west. Uncommon.
**DESCRIPTION** A very small crab (carapace to 2cm
wide) and easy to overlook. Mottled brown and
hairy on all body parts. Once found this is a very
distinctive crab.

Underwater

# Furrowed Crab
*Xantho incisus*

**HABITAT** Lower shore.
**WHERE TO LOOK** Well hidden in rockpools, under
stones and seaweeds. Porlock Weir and west.
Uncommon.
**DESCRIPTION** Carapace (to 7cm wide) is red/brown.
Reminiscent of a small Edible Crab but the
furrowed carapace is distinctive. Heavy pincers
can be black- or brown-tipped.
Legs may be sparsely hairy
close to carapace.

Distinctive furrowed carapace

# Risso's Crab
## *Xantho pilipes*

**HABITAT** Lower shore.
**WHERE TO LOOK** Well hidden in rockpools, under stones and seaweeds. Minehead and west. Uncommon.
**DESCRIPTION** Pale red/orange and sometimes mottled with white and brown patches. Has a furrowed carapace (to 6cm wide) and is densely hairy on legs and on underside of carapace. The pincers are usually pale brown.

Densely hairy on legs

# Velvet Swimming Crab
## *Necora puber*

**HABITAT** Lower shore.
**WHERE TO LOOK** Under seaweeds, rocks and boulders. Watchet and west. Common.
**DESCRIPTION** Distinctive crab (carapace to 6cm wide) with red eyes and blue/purple marking on legs and pincers. An active predator feeding on smaller crabs, fish, molluscs and worms.

Blue/purple markings on legs

# Pennant's Swimming Crab
*Portumnus latipes*

Hind leg end segment paddle-shaped

**HABITAT** Middle and lower shore.
**WHERE TO LOOK** On sand, much of the time buried in sand and hidden from view. Often found washed up along the tideline. All beaches. Common.
**DESCRIPTION** Pale sandy-coloured crab, mottled, sometimes with white patches. The paddle-shaped end segments of the hind legs are an important diagnostic feature as is the carapace (to 2cm wide) shape, which is longer front to back than the width.

33

# South-claw Hermit Crab
*Diogenes pugilator*

**HABITAT** Lower shore.
**WHERE TO LOOK** Sand beaches and also in rockpools and amongst boulders. Minehead and west. Common.
**DESCRIPTION** A very small crab (carapace to 2cm long) nearly always found in Netted Dog Whelk shells (the one pictured here is in a Common Periwinkle shell). When facing you the leg on the right has the much bigger claw. Feeds on small live prey and also carrion.

# Common Hermit Crab
*Pagarus bernhardus*

**HABITAT** Middle and lower shore.
**WHERE TO LOOK** In rockpools and amongst rocks and boulders but also on sandy beaches. All beaches. Very common.
**DESCRIPTION** Carapace to 3cm long. Occupies a range of shells including Periwinkles and Dog Whelks (the one pictured here is in a Common Whelk shell). Larger specimens are usually in Whelk shells. When facing you the leg on the left has the much bigger claw. Feeds on small live prey and also carrion.

# Broad-clawed Porcelain Crab
*Porcellana platycheles*

**HABITAT** Lower shore.
**WHERE TO LOOK** Under rocks and boulders. Watchet and west. Common.
**DESCRIPTION** Tiny crab (carapace to 15mm wide) that is very well camouflaged by the silt-covered hairs on its carapace and legs which make it look very muddy. Can be hard to spot underneath a rock despite them often clustering together in good numbers.

Hairy carapace and legs

## Long-clawed Porcelain Crab
*Pisidia longicornis*

**HABITAT** Lower shore.
**WHERE TO LOOK** Under rocks and boulders.
St Audries Bay and west. Uncommon.
**DESCRIPTION** A very tiny (carapace to 10mm wide),
easily overlooked crab. Carapace is circular and
smooth and mottled red/brown.

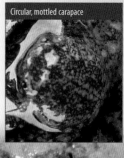
Circular, mottled carapace

## Common Prawn
*Palaemon serratus*

**HABITAT** Middle and lower shore.
**WHERE TO LOOK** In rockpools and under rocks and boulders in rockpools. All beaches.
Common.
**DESCRIPTION** A carrion feeder (to 7cm long). Transparent, but well marked with
delicate red/brown stripes and with yellow dots and bands on the body and legs.
A fast and active swimmer. The **Brown Shrimp**
***Crangon crangon*** (to 8cm long), also
occurs but is often hard
to see as they regularly
bury themselves in
the sand.

## Acorn Barnacle
*Semibalanus balanoides*

**HABITAT** Middle shore.
**WHERE TO LOOK** On rocks and boulders. All beaches. Common.
**DESCRIPTION** Ivory white, to 10mm diameter. Has six shell plates and the aperture is diamond-shaped out. As barnacles mature the shell plates may become fused making it hard to count them. A hand lens is vital to sorting them. The greyer barnacles surrounding these Acorn Barnacles are Kiwi Barnacles.

Ivory white with six shell plates

## Volcano Barnacle
*Balanus perforatus*

**HABITAT** Lower shore.
**WHERE TO LOOK** On rocks and boulders. All beaches. Common.
**DESCRIPTION** Dirty white, sometimes slightly purple. A very substantial barnacle (to 3cm diameter) and, as the name implies, volcano-shaped with an oval aperture. Has six shell plates.

Volcano-shaped with oval aperture

# Castle Barnacle
*Balanus crenatus*

**HABITAT** Lower shore.
**WHERE TO LOOK** On rocks and boulders. All beaches. Common.
**DESCRIPTION** Creamy white and rather shiny, to 20mm diameter. Substantial and with a steep-sided profile. Has six shell plates. The aperture looks castellated, is slightly offset giving this barnacle a leaning look.

Creamy white with steep-sided profile

# Kiwi Barnacle
*Elminius modestus*

**HABITAT** Upper and middle shore.
**WHERE TO LOOK** On rocks and boulders. All beaches Very common.
**DESCRIPTION** Grey white, to 10mm diameter. Introduced from New Zealand, this is the commonest small barnacle found along our coast. Has four shell plates and the aperture is diamond-shaped.

Grey white with four shell plates

# Bristly Bryozoan
*Flustrellidra hispida*

**HABITAT** Lower shore.
**WHERE TO LOOK** Most often on the fronds of
Serrated Wrack but does occur on a variety of
other seaweeds. Minehead and west. Uncommon.
**DESCRIPTION** A brownish-purple and bristly
bryozoan. Forms elongated, gelatinous patches
up to 2–3mm thick and up to 50mm across.

Bristly and gelatinous patches

# Common Sea Mat
*Membranipora membranacea*

**HABITAT** Middle and lower shore.
**WHERE TO LOOK** On rocks and seaweed. All beaches.
Very common.
**DESCRIPTION** A dirty white bryozoan. Forms large
and often elongated patches, up to 30cm long, on
the fronds of the larger brown and red seaweeds.
The individual cells, zooids, are elongated
rectangles.

Cells are elongated rectangular

# Frosty Sea Mat
*Electra pilosa*

**HABITAT** Middle and lower shore.
**WHERE TO LOOK** On seaweed fronds, sometimes on
  rocks. Watchet and west. Very common.
**DESCRIPTION** A dirty white bryozoan. Forms colonies
  that are roughly snowflake-shaped in patches up to
  15cm long. Colonies that are growing close
  together may join up as they grow, but will still
  retain the ragged star-shapes on the
  outer edges. The individual
  cells, zooids, are
  round/oval

Individual cells are round/oval

# White Cup Bryozoan
*Disporella hispida*

**HABITAT** Lower shore.
**WHERE TO LOOK** On and under rocks and boulders,
  sometimes on the holdfasts and stems of the
  larger seaweeds and also shells. St Audries Bay and
  west. Common.
**DESCRIPTION** A distinctive hard, white, spiky
  colonial animal. Young colonies (pictured below)
  are cup-shaped, with a raised edge. Mature
  colonies (up to 20mm diameter)
  become domed.

Mature colony

## Pale Orange Mat
*Alcyonidium gelatinosum*

Zooids

**HABITAT** Lower shore.
**WHERE TO LOOK** On rocks and boulders. Often on the undersides of boulders and sometimes on the stems of larger seaweeds. All beaches. Common.
**DESCRIPTION** Pale orange/brown, soft and gelatinous bryozoan. Can form irregular-shaped patches up to 15cm across. The zooids are large and easy to see.

## Sea Chervil or Finger Bryozoan
*Alcyonidium diaphanum*

Broken-off piece

**HABITAT** Lower shore.
**WHERE TO LOOK** On rocks and boulders. Minehead and west. Common.
**DESCRIPTION** Dark honey-coloured bryozoan (up to 30cm long), tough and rubbery. Young specimens can be finger-shaped and finger-sized, older specimens may be irregularly branched and knobbly. Small broken-off pieces are most likely to be found along tideline.

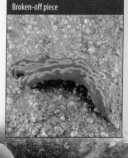

## Pale Orange Pink Mat
*Cryptosula pallasiana*

**HABITAT** Lower shore.
**WHERE TO LOOK** On rocks and also kelp holdfasts. Minehead and west. Common.
**DESCRIPTION** Pinky-orange bryozoan. Forms roughly circular patches usually 3–4cm across but can be bigger. The zooids are easily seen with the naked eye.

Zooids

## Pink Mat
*Schizomavella linearis*

**HABITAT** Lower shore.
**WHERE TO LOOK** Often found on the undersides of boulders but can also be on rock faces and holdfasts of kelp. Minehead and west. Common.
**DESCRIPTION** Pinky-red bryozoan. Forms roughly circular patches usually 3–4cm across but can be bigger. The edges of the colony are paler than the centre.

Edge of colony paler than centre

# Hornwrack
*Flustra foliacea*

**HABITAT** Lower shore.
**WHERE TO LOOK** Grows anchored to rock and
boulders, mostly subtidal. Most often found
washed up along the tideline. Blue Anchor and
west. Common.
**DESCRIPTION** Brown/grey to pale fawn bryozoan to
20cm long. Has very thin flat leathery branching
fronds with rounded tips.

# Grey Chiton
*Lepidochitona cinerea*

**HABITAT** Lower shore.
**WHERE TO LOOK** On and under rocks and stones,
sometimes on the larger seaweeds. All beaches.
Common.
**DESCRIPTION** Dark yellow/brown with some darker
markings mottling the shell. Up to 25mm long.
The shells of all our chitons are articulated by
eight plates. Also known Coat
of Mail Shells

Articulated by eight plates

# Northern Red Chiton
*Tonicella rubra*

**HABITAT** Lower shore.
**WHERE TO LOOK** In rockpools, on and under rocks.
Porlock and west. Uncommon.
**DESCRIPTION** This species (up to 20mm long) is very
variable in colour. Often red/pink but can be
greenish-grey to almost black. Grazes mainly on
algae but also on bryozoans and other colonial
mat animals.

Greenish coloured form

43

# Spiny Chiton
*Acanthochitona crinita*

**HABITAT** Lower shore.
**WHERE TO LOOK** In rockpools, on and under rocks.
Dunster Beach and west. Common.
**DESCRIPTION** A robust-looking shell to 30mm long.
Mottled pale green and brown and with a distinct
fringe of clumped bristles and spines. Grazes
mainly on algae but also on bryozoans and
other colonial mat animals.

Typically greenish-grey in colour

# Slit Limpet
*Emarginula fissura*

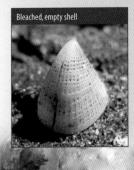
Bleached, empty shell

**HABITAT** Lower shore.
**WHERE TO LOOK** On and under rocks and boulders. Watchet and west. Uncommon.
**DESCRIPTION** Dull white-yellow or grey. This small limpet (up to 10mm long and 8mm high) is very easy to overlook and is most often found empty and bleached white on the high tideline at Minehead. Feeds on sponges.

# Common Limpet
*Patella vulgata*

**HABITAT** Upper and middle shore.
**WHERE TO LOOK** On rocks and boulders and in rockpools. All rocky beaches. Very common.
**DESCRIPTION** Neat conical shape, up to 60mm long. Grey-white in colour. Inside of shell is green/grey. The sole of the foot is yellow/orange. Can live up to 15 years. Grazes on algae. The **China Limpet** *Patella ulyssiponensis* also occurs on the lower shore and is very similar to the Common Limpet.

# Black-footed Limpet
*Patella depressa*

**HABITAT** Lower shore.
**WHERE TO LOOK** On rocks and boulders. Porlock and west. Uncommon.
**DESCRIPTION** A grey-white limpet (to 30mm long) with broad radiating bands of
brown/green. Flattened conical with radiating ridges and the apex well offset. The
sole of the foot is black. Grazes on algae.

# White Tortoiseshell Limpet
*Tectura virginea*

**HABITAT** Lower shore.
**WHERE TO LOOK** Under rocks and in rockpools with Pink Paint Weeds. Watchet and
west. Uncommon.
**DESCRIPTION** A very small limpet (to 10mm long), often with encrusting pink seaweed
on the shell. Shiny and white/grey with pink/brown rays radiating from the spire.
Typical limpet shape and with an offset spire. Feeds on red
seaweeds particularly Pink Paint Weeds.

# Blue-rayed Limpet
*Hecion pellucidum*

**HABITAT** Lower shore and subtidal.
**WHERE TO LOOK** On kelp stems and holdfasts which they partly burrow into as they feed. Porlock and west. Uncommon.
**DESCRIPTION** A neat, oval, shiny amber-coloured shell (up to 15mm long) with bright blue lines and spots. The lines and spots fade quickly on empty shells found washed ashore. Feeds on the stems of large seaweeds such as Oarweed

Empty shell

# Slipper Limpet
*Creipidula fornicata*

**HABITAT** Lower shore.
**WHERE TO LOOK** On sand and mud beaches. Watchet and west. Uncommon.
**DESCRIPTION** Beige/brown shell, to 50mm long. Oval and smooth. Often found with several attached together. Introduced from North America.

Underside of empty shell

# Thick Top Shell
*Osilinus lineatus*

**HABITAT** Middle and lower shore.
**WHERE TO LOOK** In rockpools and on rocks and
boulders. Watchet and west. Very common.
**DESCRIPTION** Heavy-shelled mollusc (up to 30mm
high) with dense, dull purple- brown zigzag
pattern. Shell opening has an obvious small
tooth-like projection. Top of shell often
worn and showing mother of
pearl. Feeds on algae.

Underside

# Painted Top Shell
*Calliostoma zizyphinum*

**HABITAT** Lower shore.
**WHERE TO LOOK** In rockpools and under seaweed.
Minehead and west. Uncommon.
**DESCRIPTION** Very distinctive, steep-pointed shell
(to 30mm high) with bright pinky-purple and red
streaks on a pale brown-white background. Feeds
on algae.

Distinctive steep-pointed shell

## Grey Top Shell
*Gibbula cineraria*

**HABITAT** Lower shore.
**WHERE TO LOOK** In rockpools and under rocks and seaweed. Watchet and west. Common.
**DESCRIPTION** Shell (up to 30mm high) has a neat, pale red-brown chequered pattern. Empty and worn shells found along the tideline often show large areas of mother of pearl. Feeds on algae.

## Flat Top Shell
*Gibbula umbilicalis*

Underside

**HABITAT** Middle and lower shore.
**WHERE TO LOOK** In rockpools and under seaweed. Kilve and west. Very common.
**DESCRIPTION** Slightly squashed looking, flat-topped shell (to 20mm high) with bright purple bands over a grey-green background. Has a distinctive large umbilicus (hole) in the centre of the underside of the shell. Feeds on algae.

# Pheasant Shell
*Tricolia pullus*

Shell with pattern of swirling lines

**HABITAT** Lower shore.
**WHERE TO LOOK** Usually amongst red seaweeds. Most often found as an empty shell washed-up along the upper tideline. Minehead and west. Uncommon.
**DESCRIPTION** A tiny shell (usually 4–5mm long, but can be up to 8mm) with a pattern of red and yellow swirling lines.

# Mud Snail or Laver Spire Shell
*Hydrobia ulvae*

Bleached, empty shell

**HABITAT** Middle and lower shore.
**WHERE TO LOOK** Often found moving across the surface of muddy shores as the tide drops. All beaches with sand and mud, particularly Burnham, Berrow and Brean beaches. Very common.
**DESCRIPTION** A brown-grey, steep-sided conical shell. Very small (to 6mm long) and easy to overlook. Often found empty and bleached, higher up the shore.

# Common Periwinkle
*Littorina littorea*

A lighter-coloured specimen

**HABITAT** Upper, middle and lower shore.
**WHERE TO LOOK** On rocks and boulders. All beaches. Very common.
**DESCRIPTION** A heavy rounded shell to 30mm high, smooth and pointed and coloured dark brown-black, often with thin white bands. Top of spire sometimes lighter in colour.

# Flat Periwinkle
*Littorina obtusata*

Brown-black specimen

**HABITAT** Middle and lower shore.
**WHERE TO LOOK** On rocks and boulders, usually amongst brown seaweeds. All beaches. Very common.
**DESCRIPTION** A small flattened shell to 10mm high. Often yellow but dull red, olive green and brown-black ones also found. Their small yellow eggs, in a flattened gelatinous blob (below left) can often be found glued onto Serrated Wrack and other seaweeds.

# Rough Periwinkle
*Littorina saxatilis*

Ridged whorls

**HABITAT** Upper and middle shore.
**WHERE TO LOOK** Often found in rock crevices and also amongst seaweeds. All beaches. Very common.
**DESCRIPTION** A solid rounded shell (to 17mm high) with a short pointed spire and ridged whorls making it rough to the touch. Brown but can be orange. The tiny **Small Periwinkle** (4–6mm high) *Melarhaphe neritoides* also occurs on the upper shore usually in crevices in rocks.

# Spotted Cowrie
*Trivia monacha*

Two to three black spots on shell

**HABITAT** Lower shore.
**WHERE TO LOOK** Underside of rock overhangs and in hollows between large boulders. Hurlstone Point and west. Uncommon.
**DESCRIPTION** Distinctive glossy pinkish shell (to 12mm long) with two to three black spots and with a ribbed surface. When moving the flesh-coloured foot wraps around the shell. Feeds on ascidians.

# Arctic Cowrie
*Trivia arctica*

Unspotted ridged surface

**HABITAT** Lower shore and subtidal.
**WHERE TO LOOK** Underside of rock overhangs and in hollows between large boulders. Hurlestone Point and west. Uncommon.
**DESCRIPTION** Distinctive glossy pinkish shell (to 10mm long) with a ribbed surface. When moving the flesh-coloured foot wraps around the shell. Feeds on ascidians.

# Sting Winkle
*Ocenebra erinacea*

Heavily sculpted shells

**HABITAT** Lower shore.
**WHERE TO LOOK** In rockpools and amongst seaweeds. Watchet and west. Very common.
**DESCRIPTION** A distinctive pointed and sculpted yellow-white shell to 50mm long. Feeds on bivalves and barnacles.

# Common Whelk
*Buccinum undatum*

**HABITAT** Lower shore.
**WHERE TO LOOK** Often found washed up as an empty shell along the tideline, as are their white spongy eggs. All beaches. Uncommon.
**DESCRIPTION** A distinctive large yellow-white shell to 110mm long. The shell is much thinner than in the Dog Whelk. Small specimens can be told from Dog Whelks by this feature. Feeds on tube-dwelling worms, cockles, other bivalves and carrion.

White spongy eggs

# Dog Whelk
*Nucella lapillus*

**HABITAT** Middle and lower shore.
**WHERE TO LOOK** On rocks and boulders. All beaches. Common.
**DESCRIPTION** A heavy, pointed, often rather rough shell to 30mm long. Very variable in colour including white, pink and black and also with brown stripes. In spring and early summer, large clusters of their distinctive yellowish eggs can be found on rocks and boulders on the lower shore. Feeds on barnacles and mussels.

Dog Whelk egg-laying

## Netted Dog Whelk
*Nassarius reticulatus*

**HABITAT** Middle and lower shore.
**WHERE TO LOOK** In rockpools and amongst
seaweeds. Blue Anchor and west. Common.
**DESCRIPTION** A robust, fawn to brown-grey shell (to
30mm long) with well-marked grooves and ribs.
Their pinky-white eggs, enclosed in a translucent
flattened capsule, can often be found glued to
rocks or stems of seaweeds. The
empty shell is often occupied
by hermit crabs.

Eggs enclosed in translucent capsule

## White Sea Slug
*Goniodoris nodosa*

**HABITAT** Lower shore.
**WHERE TO LOOK** In rockpools and on rocks and
boulders. Watchet and west. Uncommon.
**DESCRIPTION** Translucent white and very small sea
slug to 25mm long. If startled and hunched-up it
looks like a shiny white oval blob. Feeds on
bryozoans and ascidians.

Hunches up if startled

# Sea Hare
*Aplysia punctata*

Eggs are laid in distinctive pink ribbons

**HABITAT** Lower shore.
**WHERE TO LOOK** In rockpools and amongst seaweeds. Hurlestone Point and west. Sometimes very common.
**DESCRIPTION** Distinctive sea slug (to 100mm long), reddish-brown with scattered green/white dots. Can gather in hundreds in rockpools in summer during breeding season. They lay their eggs in distinctive, pink sticky ribbons. Feeds on seaweeds.

# Green Sea Slug
*Elysia viridis*

Feeds on green seaweeds

**HABITAT** Lower shore.
**WHERE TO LOOK** In rockpools and amongst seaweeds. Porlock Weir and west. Uncommon.
**DESCRIPTION** A sea slug to 40mm long, green with very small blue spots. It can be very hard to spot amongst green seaweeds upon which it feeds.

# Dog Cockle
*Glycymeris glycymeris*

**HABITAT** Lower shore.
**WHERE TO LOOK** Lives buried in the sand. Most likely to be found washed up on the tideline. Blue Anchor and west. Uncommon.
**DESCRIPTION** Smooth yellowish shell (to 60mm wide) with red/brown, concentric zigzag pattern. A filter-feeder on plankton.

# Common Mussel
*Mytilus edulis*

**HABITAT** Lower shore.
**WHERE TO LOOK** Attached to rocks and in crevices between rocks and boulders. Kilve and west. Common.
**DESCRIPTION** Distinctive blue-purple/black bivalve to 60mm long. Uncommon as live animal but shells commonly washed up on tideline. A filter-feeder on microscopic plankton.

Washed-up shell on tideline

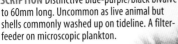

# Saddle Oyster
*Anomia ephippium*

**HABITAT** Lower shore.
**WHERE TO LOOK** On undersides of rocks and boulders. St Audries Bay and west. Common.
**DESCRIPTION** White/grey or pale golden-brown shiny bivalve (to 50mm wide), closely attached to rocks. A filter-feeder on plankton.

Closely attached to rocks

# Variegated Scallop
*Chlamys varia*

**HABITAT** Lower shore.
**WHERE TO LOOK** On rocks and boulders, but most likely to be found washed up on the tideline. Minehead and west. Scarce.
**DESCRIPTION** Dark red/brown bivalve (to 60mm long), mottled with white. Appears to only have one "ear", part of the hinge between the two shells. A filter-feeder on plankton. It can be found washed up in very large numbers along the tideline on the Welsh coast opposite.

Red/brown shell mottled with white

# Common Oyster
*Ostrea edulis*

**HABITAT** Lower shore.
**WHERE TO LOOK** Uncommon as a living animal but shells often found along the tideline. Blue Anchor and west. Common.
**DESCRIPTION** Grey white bivalve with a rough circular to oval shell to 90mm wide. A filter-feeder on plankton.

# Pacific or Portuguese Oyster
*Crassostrea gigas*

**HABITAT** Lower shore and subtidal.
**WHERE TO LOOK** On rocks and boulders. All beaches. Uncommon.
**DESCRIPTION** Roughly oval shell (to 20cm long) with irregular flaky ridges, tightly cemented to rocks. Grey white and often with purple-brown bands. A filter-feeder on microscopic plankton. A non-native species that has spread from oyster farms.

# Pullet Carpet Shell
*Venerupis senegalensis*

**HABITAT** Lower shore in sand.
**WHERE TO LOOK** In sand, but most often found washed up along the tideline. Dunster Beach and west. Common.
**DESCRIPTION** A heavy elongated oval shell (to 50mm wide) with several concentric ridges and numerous very narrow radiating ridges. Fawn with some brown markings. A filter-feeder on plankton.

# Striped Venus Shell
*Chamelea gallina*

**HABITAT** Lower shore.
**WHERE TO LOOK** In sand. Minehead and west. Scarce.
**DESCRIPTION** Robust shell (to 40mm wide) with concentric ridges. Pale fawn and with radiating broken brown rays. A filter-feeder on plankton.

Concentric ridges and brown rays

# Baltic Tellin
*Macoma balthica*

**HABITAT** Lower shore.
**WHERE TO LOOK** In sand, but most often found dead and washed up along the tideline. All beaches, but often in vast numbers washed up at Burnham, Berrow and Brean. Very common.
**DESCRIPTION** Shell, to 25mm wide, is rounded oval and thick. Colour varies enormously but is often banded with pale yellow/red or pale black/purple. A filter-feeder on plankton.

Rounded oval shape

# Common Cockle
*Cerastoderma edule*

**HABITAT** Lower shore.
**WHERE TO LOOK** In sand, but most often found dead and washed up along the tideline. All beaches, but very common on the beach between Blue Anchor and Dunster Beach. Common.
**DESCRIPTION** Dirty white bivalve (to 40mm wide) with 24 or so rough radiating ridges. Classic cockle shape is easy to identify. A filter-feeder on plankton.

Classic cockle shape

# Common Otter Shell
*Lutraria lutraria*

**HABITAT** Lower shore.
**WHERE TO LOOK** In sand, but most often found washed up along the tideline. All beaches. Uncommon.
**DESCRIPTION** A large and robust shell to 12cm long. Pale brown with a darker brown, flaky covering around outer edges of shell. A filter-feeder on plankton.

# Banded Wedge Shell
*Donax vittatus*

**HABITAT** Lower shore.
**WHERE TO LOOK** In sand, but most often found washed up along the tideline. All beaches. Common.
**DESCRIPTION** Shell (to 35mm long) is elongated and slightly triangular. Blue/purple concentric bands on a fawn background. A filter-feeder on plankton.

Often found washed up on tideline

# Common Piddock
*Pholas dactylus*

**HABITAT** Lower shore.
**WHERE TO LOOK** In bore holes in soft rock. All beaches. Common.
**DESCRIPTION** Dirty white bivalve to 10cm long. The outside of shell has ridges covered with very small, hard tooth-like structures, which enable it to bore into soft rock. Lives in the hole it bores with only a small part of the shell protruding. A filter-feeder on plankton.

Ridges covered in tooth-like structures

# Common Cuttlefish
*Sepia officinalis*

**HABITAT** Lower shore.
**WHERE TO LOOK** Cuttlefish egg-clusters are found on seaweeds at the lowest point of the tide. Blue Anchor and west. Minehead appears to be an important breeding area. Shells may be found along tideline of all Somerset beaches. Uncommon.
**DESCRIPTION** Adult Cuttlefish (to 35cm long) have an oval body with side fins, large eyes, and short tentacles. Usually mottled streaky brown but can change colour rapidly to match background. Eggs are dark brown/purple and 10–12mm diameter. They are laid in bunches and attached to seaweeds.

# Common Squid
*Loligo vulgaris*

**HABITAT** Lower shore.
**WHERE TO LOOK** In rockpools. Minehead and west. Scarce.
**DESCRIPTION** Adult (to 70cm long) has an elongated shape and large eyes. May be almost transparent but can rapidly change colour to match background. Squid eggs (to 7cm long) are elongated, white and jelly-like. They can be found attached to seaweeds at the very lowest point of the tide and may, very occasionally, be washed ashore.

Can change colour rapidly

# Common Starfish
*Asterias rubens*

**HABITAT** Lower shore.
**WHERE TO LOOK** In rockpools and under large stones. Minehead and west. Uncommon.
**DESCRIPTION** A pale brown/orange starfish (to 30cm diameter including arms) with short white spines in lines down the five, very stout, arms. They feed on molluscs and the larger marine worms.

Detail showing underside of an arm

# Common Sunstar
*Crossaster papposus*

Large central disk and 11–13 arms

**HABITAT** Lower shore.
**WHERE TO LOOK** In rockpools and amongst
boulders. Minehead and west. Common.
**DESCRIPTION** Unmistakable bright orange/red
starfish (to 20cm diameter including arms) with a
large central disk and 11–13 relatively short arms.
May move quite rapidly on being disturbed.

# Common Brittlestar
*Ophiothrix fragilis*

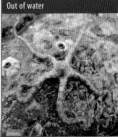
Out of water

**HABITAT** Lower shore.
**WHERE TO LOOK** Under stones and amongst
boulders. Minehead and west. Common.
**DESCRIPTION** A brittlestar (diameter including arms
up to 80mm, central disk to 10mm) with a very
spiny body and arms. Often orange and yellow
banded along the arms, but can be much duller
fawn and brown banded.

# Small Brittlestar
*Amphipholis squamata*

**HABITAT** Middle and lower shore.
**WHERE TO LOOK** In rockpools and under stones. Often half buried in sand and gravel and very difficult to spot. St Audries Bay and west. Common.
**DESCRIPTION** A very small grey-black brittlestar (to 15mm diameter, including arms) with the central disk only 3–4mm.

Small central disk 3–4mm diameter

# Baked Bean Ascidian
*Dendrodoa grossularia*

**HABITAT** Lower shore.
**WHERE TO LOOK** On rocks and boulders, sometimes on the stem base of the larger seaweeds and also on mollusc shells. All beaches. Common.
**DESCRIPTION** Individuals are orange/red to 10mm high. They have two orifices, one for sucking in food particles and one to eject waste matter. They can be found as solitary individuals but are usually in groups. Intertidal colonies may cover only a few square centimetres, but cover meters subtidal.

Sometimes found on seaweed stems

# Star Ascidian
*Botryllus schlosseri*

**HABITAT** Lower shore.
**WHERE TO LOOK** Under rock overhangs and large
  boulders. Porlock Weir and west. Uncommon.
**DESCRIPTION** Forms patches to 80mm across.
  Distinctive groups of between 6–13 beige/white,
  sometimes yellow, zooids forming star-shaped
  clusters in a gelatinous body that is transparent
  grey or purple.

Zooids sometimes yellow

# Small-spotted Catshark (egg case)
*Scyliorhinus canicula*

**HABITAT** Adults far out at sea. Eggs are laid well
offshore in seaweed beds.
**WHERE TO LOOK** Egg cases washed up along the
  tideline. All beaches. Common.
**DESCRIPTION** Small-spotted Catsharks grow to 75cm.
  They may lay up to 100 eggs per annum. Egg case
  (to 5cm long) is golden brown when fresh and may
  turn black as it dries out. Has long curling tendrils
  which secure it to seaweeds etc.
  **Nursehound *Scyliorhinus
  stellaris*** egg case is
  similar but up
  to 8cm long.

Nurse Hound egg case

## Blonde Ray (egg case)
*Raja brachyura*

**HABITAT** Adults far out at sea. Eggs are laid well offshore in seaweed beds.

**WHERE TO LOOK** Egg cases washed up along the tideline. All beaches. Uncommon.

**DESCRIPTION** Blonde Rays grow to 130cm. The large black egg case (to 14cm long including horns) has long horns at each corner, a leathery texture and will fill the palm of your hand.

Large egg case to 14cm long

## Thornback Ray (egg case)
*Raja clavata*

**HABITAT** Adults far out at sea. Eggs are laid well offshore in seaweed beds.

**WHERE TO LOOK** Egg cases washed up along the tideline. All beaches. Common.

**DESCRIPTION** Thornback Rays grow to 90cm. The egg case (to 9cm long, including horns) is golden brown when fresh but turns black as they dry out. It is rather square and has long points at each corner.

Fresher egg case turning black

## Shore Clingfish
*Lepadogaster lepadogaster*

**HABITAT** Lower shore.
**WHERE TO LOOK** In rockpools, under rocks and
amongst boulders. Minehead and west.
Uncommon.
**DESCRIPTION** A distinctive, mottled pinky-yellow/
orange fish (to 8cm long) which tapers rapidly
towards the tail. It has a very broad head with two
blue spots behind the eyes.

Two blue spots behind the eyes

## Worm Pipefish
*Nerophis lumbriciformis*

**HABITAT** Lower shore.
**WHERE TO LOOK** In rockpools and under stones. Very
easy to overlook. Minehead and west. Uncommon.
**DESCRIPTION** Very slender, worm-like, yellow/brown
fish (to 15cm long) with a short upturned nose and
a stiff body. Has a single small dorsal fin about half
way along the body and no tail fin. The similar, but
larger **Snake Pipefish *Entelurus
aequoreus*** and **Great
Pipefish *Syngnathus
acus*** also occur
along this
coast.

Short upturned nose

# Lesser Sand Eel
*Ammodytes tobianus*

**HABITAT** Lower shore.
**WHERE TO LOOK** In rockpools. Minehead and west. Common.
**DESCRIPTION** A distinctive, very slender, elongated silvery fish (to 20cm long) with grey/yellow back and white/silver underside. Swims very rapidly and likely to be several swimming together. Will dive down and burrow into sand when alarmed. The larger **Greater Sand Eel *Hyperoplus lanceolatus*** (up to 35cm long) is also likely to be present along this coast.

# Shanny
*Lipophrys pholis*

**HABITAT** Lower shore.
**WHERE TO LOOK** In rockpools, and often found under large stones and amongst boulders. All beaches. Common.
**DESCRIPTION** Marbled brown/grey fish (to 13cm long), sometimes almost black. Has one long dorsal fin along the back with a dip in it half way along, and very large pectoral fins. Has no scales and so looks rather slimy out of water.

Looks slimy out of water

## Montagu's Blenny
*Coryphoblennius galerita*

**HABITAT** Upper and middle shore.
**WHERE TO LOOK** In rockpools and amongst stones.
  Minehead and west. Uncommon.
**DESCRIPTION** Mottled black/brown fish (to 8cm
  long) with grey strips and patches. Orange/red
  markings along top of dorsal fin. The small
  tentacles between the eyes which help to
  identify this species are not
  always clearly seen
  from above.

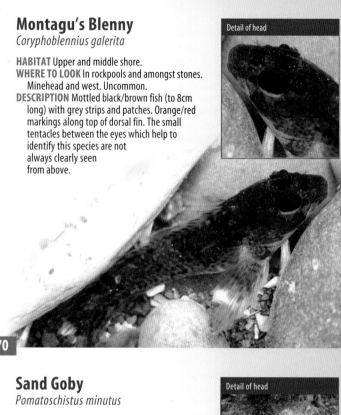

Detail of head

## Sand Goby
*Pomatoschistus minutus*

**HABITAT** Middle and lower shore.
**WHERE TO LOOK** On sand in rockpools. All beaches.
  Common.
**DESCRIPTION** Pale sandy-brown fish (to 6cm long)
  with small dark dots all over the body. The two
  dorsal fins are clearly separated. Unless they
  move they can be very hard to see when
  resting on sand.

Detail of head

# Painted Goby
*Pomatoschistus pictus*

**HABITAT** Lower shore.
**WHERE TO LOOK** In rockpools and amongst boulders. Porlock and west. Uncommon.
**DESCRIPTION** A small fish (to 5cm long), fawn with dark bands and pale spots along the sides of the body. Bands and spots of black on the dorsal fins are an important diagnostic feature for this species as other gobies do not have these.

Detail of head

# Rock Goby
*Gobius paganellus*

**HABITAT** Lower shore.
**WHERE TO LOOK** In rockpools and under rocks. Watchet and west. Uncommon.
**DESCRIPTION** Marbled brown/grey fish (to 12cm long), sometimes almost black. The two dorsal fins are barely separated and have a pale orange/yellow band along the top edge. The head is relatively large.

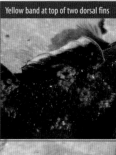
Yellow band at top of two dorsal fins

# Species index